GW01311679

MORSE CODE

Personal Data

Name: _____

Address: _____

Phone number: _____

Email: _____

What's Your Story?
Tell us your story about learning Morse code as you can inspire others to learn.
We would really love to get your review and feedback on Amazon.

"The secret of getting ahead is getting started"
– Mark Twain

MORSE CODE

"In 1825 a painter named Samuel Morse. In Washington, D.C. Morse received a letter from his father -delivered via the standard, slow-moving horse messengers of the day- as his wife suddenly had suffered a heart attack while giving birth to their third child and she died, Morse immediately left the capital and hurried home in Connecticut. By the time he arrived, however, his wife wasn't only dead, she was already buried. He was so mad about the slow messages and distraught at the death of his wife, that he realized that we need a faster communication method and invented one. He called it Morse Code."

- Morse code is named for its inventor, Samuel Morse.

- The first official message to be sent in Morse code on a united states experimental telegraph line was relayed between Washington, D.C. and Baltimore, MD on May 24 ,1844.

- Morse code converts letters and numbers into a series of dots and dashes (sometimes called dits and dahs).

- In Morse code, each dash has a duration that is three times as long as each dot. Each dot or dash within a character is followed by a period of no signal, called a space, equal in duration to the dot.

- Morse code is well suited to be communicated through sound using audio tones. It can also be communicated visually using flashing lights or eye blinks.

- Although the code is not designed to be transmitted in written format, it can be written as well.

Morse Code rules

Morse code is a system of communication that used dots (.) and dashes and dashes (-).

A dot looks like a period (.) and a dash is a long horizontal line (-).

A dot is called a (dit or di), and a dash is called (dah).

The sound of a dit (.) is like (bep) and the sound of dah is like (beeep).

Morse Code rules:-

- The smallest unit in morse code is one unit.

- A dit is equal one unit.

- A dash is equal 3 units.

- The space between dits and dashes in the same letter is equal one unit (pause).

- The space between two letters in the same word is 3 units (pause).

- The space between two words is equal 7 units (pause) also a slash can be used instead of the 7 units of pause.

- There is no distinction between upper- and lower-case letters.

- For more understanding check the next message:

"I am Happy"   in morse code is   ..   .- --   .... .- .--. .--. -.--

I   a   m   h   a   p   p   y

Also, the same sentence can be written with slashes between words as:

A ·—
B —···
C —·—·
D —··
E ·
F ··—·
G ——·
H ····
I ··

J ·———
K —·—
L ·—··
M ——
N —·
O ———
P ·——·
Q ——·—
R ·—·

S ···
T —
U ··—
V ···—
W ·——
X —··—
Y —·——
Z ——··

1	2	3	4	5
.−−−−	..−−−	...−−	....−	.....

6	7	8	9	0
−....	−−...	−−−..	−−−−.	−−−−−

(		@		)
−.−−.		.−−.−.		−.−−.−

.	?	!	'	‚
.−.−.−	..−−..	−.−.−−	.−−−−.	−−..−−

+	−	/	=	:
.−.−.	−....−	−..−.	−...−	−−−...

A

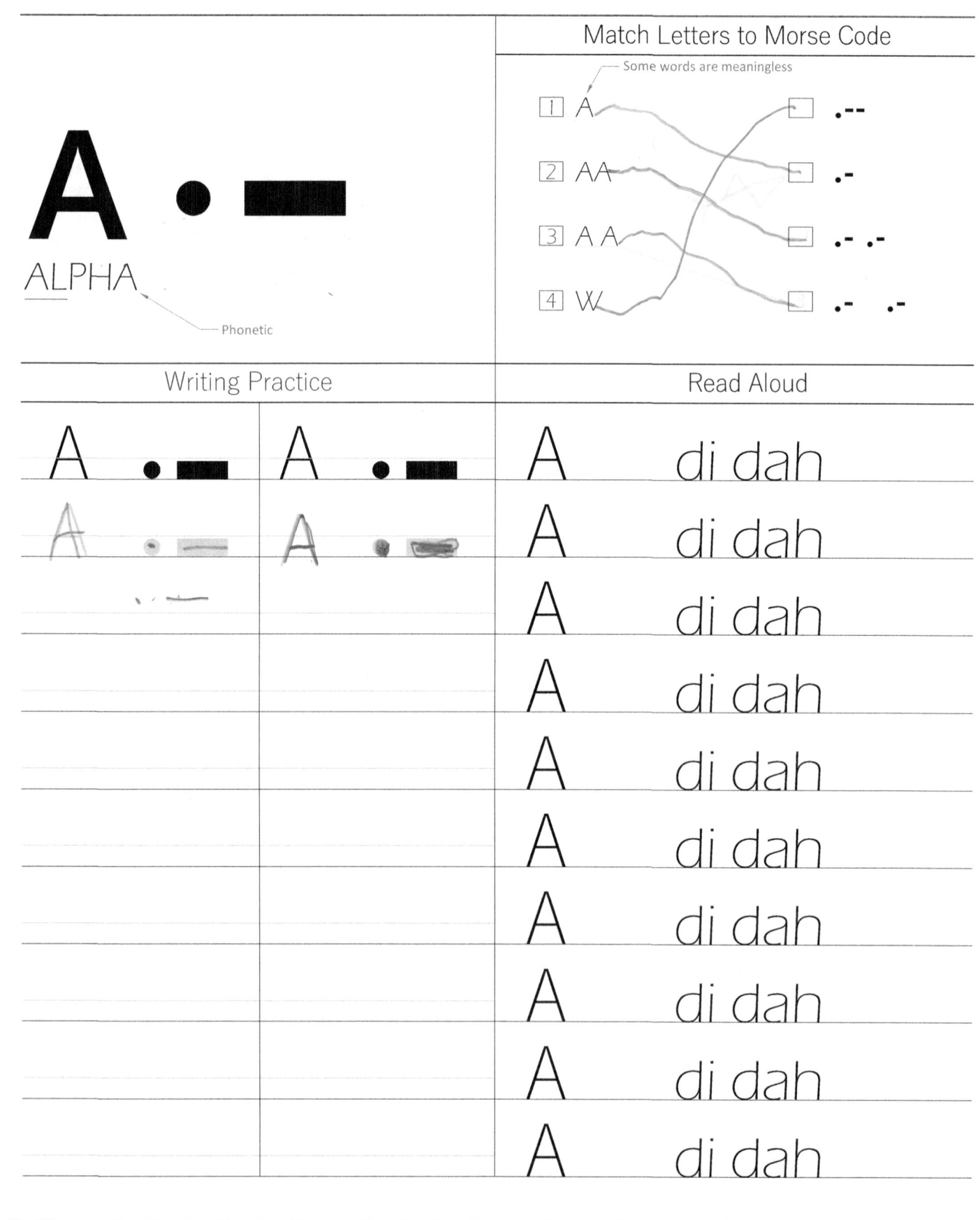

ALPHA

— Phonetic

— Some words are meaningless

1 A
2 AA
3 A A
4 W

Writing Practice

Read Aloud

A   di dah
A   di dah
A   di dah
A   di dah
A   di dah
A   di dah
A   di dah
A   di dah
A   di dah
A   di dah

Learned A B C D E F G H I J K L M N O P Q R S T U V W X Y Z

T ▬

TANGO

1. T
2. M
3. TF
4. T T

- -
- - -
- -
-

Writing Practice		Read Aloud
T	AT ·▬▬	T dah
T ▬	AT ·▬ ▬	T dah
		T dah
		T dah
		T dah
		AT di dah dah
		AT di dah dah
		AT di dah dah
		AT di dah dah
		AT di dah dah

Learned

A B C D E F G H I J K L M N O P Q R S T U V W X Y Z

A ·▬   T ▬

E •

ECHO

1. E
2. EE
3. ET
4. A

.-
. .
.
.-

Writing Practice

E •
E •

TEA ▬ • • ▬
TEA ▬ • • ▬

Read Aloud

E	dit
E	dit
E	dit
E	dit
E	dit
Tea	dah di di dah
Tea	dah di di dah
Tea	dah di di dah
Tea	dah di di dah
Tea	dah di di dah

M — —

MIKE

1 M
2 TT
3 MM
4 G

— —
— —.
— —
— — — —

Writing Practice

M — — MEET — — . . —
M — — MEET — — . . —

Read Aloud

M dah dah
M dah dah
M dah dah
M dah dah
M dah dah
Me dah dah dit
Me dah dah dit
Me dah dah dit
Me dah dah dit
Me dah dah dit

◇ Learned

A · — | B | C | D | E · | F | G | H | I | J | K | L | M — — | N | O | P | Q | R | S | T — | U | V | W | X | Y | Z

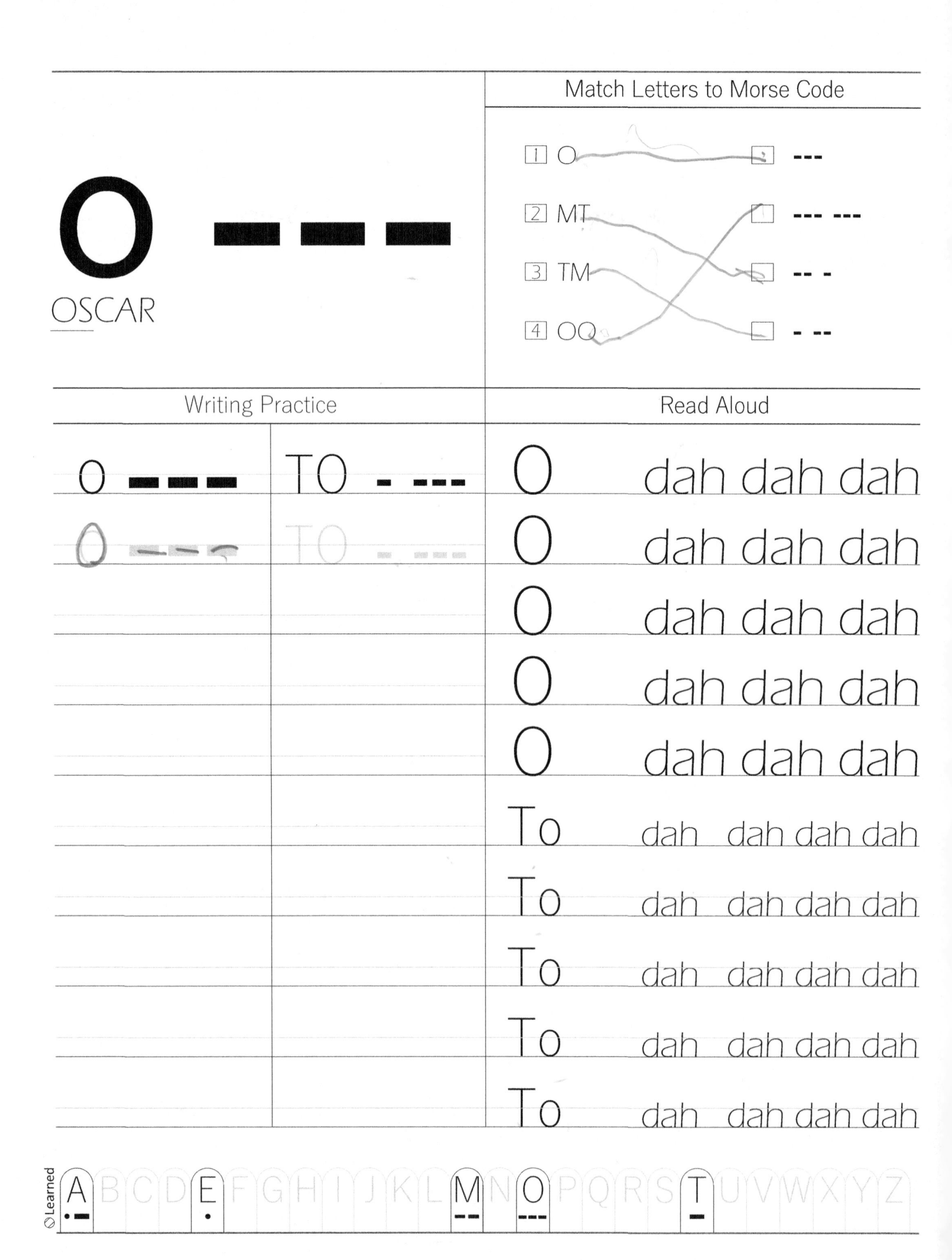

A T E M O

Match Morse words with the correct Word

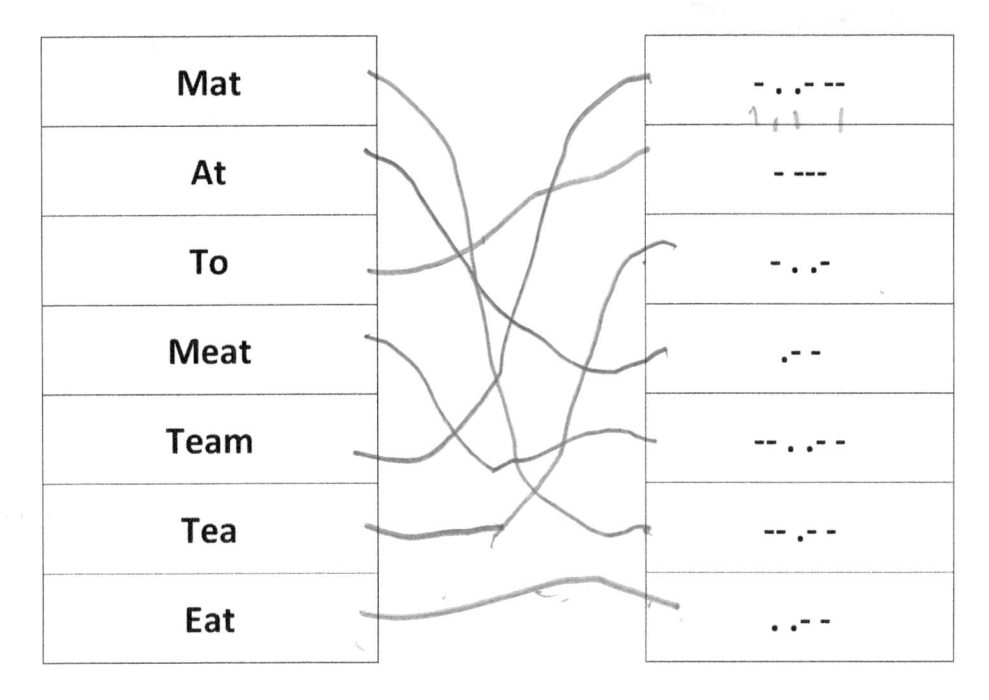

Mat	- . . - --
At	- ---
To	- . . -
Meat	. - -
Team	-- . . - -
Tea	-- . - -
Eat	. . - -

Match Morse words with the correct picture

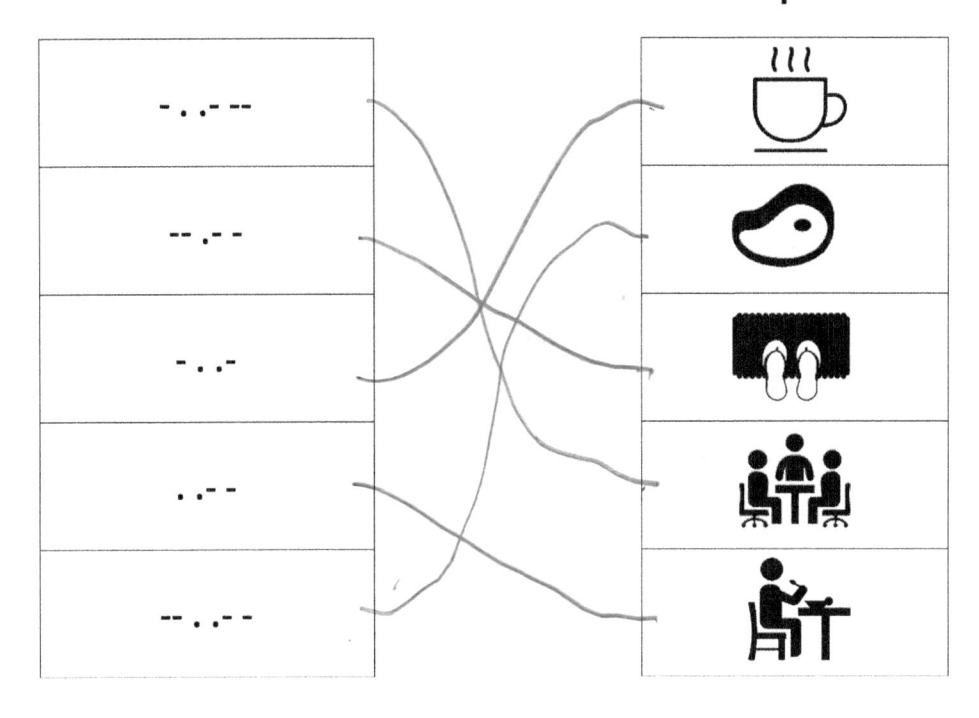

Puzzle: Decode the learned letters and extract 5 words 🔍

-.-	-.-	.--	-.	.-..	--..	-	---	--	--..	.-	-
...	.--	--	...-	.	--.	.-	-.--	---	--.-	..	.--.
.---	.--	.	.--	.-	.	.-	-.	.	--.	-.-.	---
.---	...	.-.	.--	.-	-...	-	.--	.-.	..-.	-..	-.-.
.-	.--	.-	.--	---	-.	.-.	--	.	.-	---	...
-	.-	.	.-	--	..-.	-.	.-	--..	---	-..	-.-.
.	--..	-.-.	--	.-	--.-	-.--	-	-.--	.---	.---	.-
-.-.	....	-	-	--	-...	--..	-	.	.-	.-.	-.-.

Translate text to Morse code

Text	Morse
EAT	· ·.— —
To	— ——— —
At	·— —
Team	— · ·— ——
Me	—— ·
Tea	— ···— · —
Atom	·— — ——— ——
MEAT	—— ·· ·— —

N — ●

NOVEMBER

1. N   ●—
2. A   ●— ●—
3. AA   —●
4. NN   —● —●

Writing Practice

N —● ●	NO —● ———	
N — ●	NO —● ——	

Read Aloud

N   dah dit
N   dah dit
N   dah dit
N   dah dit
N   dah dit

Extract the letter "N" and circle it:-(4)

●— ●—● ● / —●—— ——— ●●— /

●●—● ●—● ●●● / ——— —● /

●●● ●●— (—●) —●● ● ●— —●—— /

●— ●●—● — ● ●—● —●— ● ●●● ——— ——— —●

A ●—   E ●   M —— N —● O ———   T —

S ● ● ●

SIERRA

1. S
2. SS
3. S S
4. EEE

... ...
... ...
... ...
...

Writing Practice

S ● ● ●	SOS ... --- ...
S ● ● ●	SOS ... --- ...
S ...	SISTER

Read Aloud

S	di di dit
S	di di dit
S	di di dit
S	di di dit
S	di di dit

Extract the letter "S" and circle it :- (3)

.. / .- -- /

... .- - .. ..-. .. . -.. /

.- .. - .... . / - .... . /

... . .- . ... .- .. -.- .

I

INDIA

● ●

1	I			..
2	IE			... .
3	S			...
4	EI			...

Writing Practice		Read Aloud

I ● ●   Tie ‑ ●● ●   i   di dit

I ● ●   Tie ‑ ●● ●   i   di dit

i   di dit

i   di dit

i   di dit

Extract the letter "I" and circle it:‑(2)

.. / .‑ ‑‑ / ‑‑. ‑‑‑ .. ‑. ‑‑. /

‑ ‑‑‑ / ‑ ●●●● ● /

●‑●● ●● ‑●●● ●‑● ●‑ ●‑● ‑●‑‑

R .−.

ROMEO

1 R — .-.
2 EN — . - .
3 AE — . .-
4 ETE — . - .

Writing Practice

R .−. RAT .-. .- -

Read Aloud

R	di dah dit
R	di dah dit
R	di dah dit
R	di dah dit
R	di dah dit

Extract the letter "R" and circle it:-(3)

.- .-. . / -.-- --- ..- /

..-. . .-. . . / --- -. /

... . .- -. . .-.. / -.--- /

.- ..-. -. . . .-. . --- --- .

○Learned A B C D E F G H I J K L M N O P Q R S T U V W X Y Z

H

H ••••

HOTEL

1 H	•• •••
2 IS	•• ••
3 SI	••••
4 II	••• ••

Writing Practice

H •••• HAT •••• • - -

H •••• HAT •••• • - -

H •••• HAIR •••• • - - •

Read Aloud

H di di di dit

H di di di dit

H di di di dit

H di di di dit

H di di di dit

Extract the letter "H" and circle it:-(3)

•••• --- •-- / -- •- -• -•-- /

-•-• •••• •• •-•• -•• •-• • -• /

-•• --- / -•-- --- •-- /

•••• •- ••- •

A •- E • H •••• I •• M -- N -• O --- R •-• S ••• T -

A E M O T N S I R H

Match Morse words with the correct Word

Ant		-- .- -.
Arm		.- -. -
Man		... -- .- -. -
Smart		.- -. --
Nine		- . -.
ten		.... . -.
Hen		-. .. -. .

Match Morse words with the correct picture

.- .. -.-. .	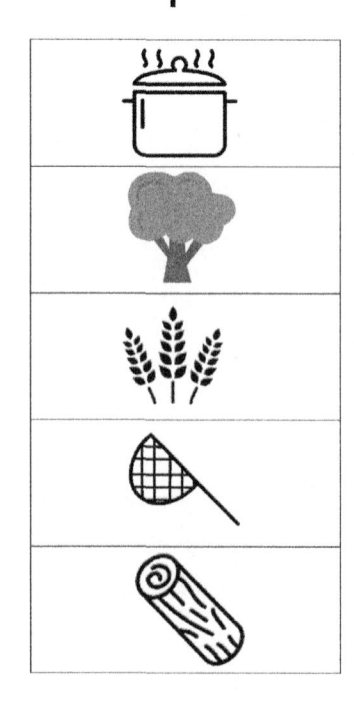
-. . -	
- .- . .	
.... --- -	
.-- --- --- -..	

Puzzle: Decode the learned letters and extract 7 words 🔍

--	-	.--	.	.-	-	-.-	--.	-.	.-	.-.	.--.
-	.	-	-.-	-	.	-.	--.-	.	-	..-.	..-.
.--	.-	-. N	.. I	-. N	. E	...	--.	-	-.--	.-	--.-
-..	.-.	-.-	-.	---	-...	...	.-	--.-	..-.	-.--	.-.
--	....	--	.-	....	...	--	.-	-.	-	...	..-.
.-	---	...-	--	--.-	.-	-.	-	-	---	.-..	-.--
-	-	..	.	--.-	...	-.-.	--.	....	.-	-	--.
.--	....	.	-.	-.	-	.	.-	--	.-.	-...	--..

Translate text to Morse code

Wood	
Tree	
Car	
Rice	
Arm	
Man	

D ▬ • •

DELTA

☐1 D		☐ -..
☐2 NE		☐ -..
☐3 TI		☐ -..
☐4 TEE		☐ -..

Writing Practice

D ▬ • •	DO -.. ---
D ▬ • •	DO -.. ---

Read Aloud

D dah di dit

D dah di dit

D dah di dit

D dah di dit

D dah di dit

Extract the letter "D" and circle it:-(4)

.-- .... . -. / -.. .. -.. /

-.-- --- ..- / --. . - /

-.. .. ..-. --- .-. -.-. . -..

Ø Learned

A B C D E F G H I J K L M N O P Q R S T U V W X Y Z
.- -... -.-. -.. . ..-. --. ....--- -.- .-.. -- -. --- .--. --.- .-. ... -

L ● ▬ ● ●

LIMA

☐1 L   ☐ .-..

☐2 RE   ☐ . -. .

☐3 ENE   ☐ . - ..

☐4 ETI   ☐ .-..

Writing Practice

L   .▬.. LOT .-.. --- -

L   LOT

Read Aloud

L   di dah di dit

L   di dah di dit

L   di dah di dit

L   di dah di dit

L   di dah di dit

Extract the letter "L" and circle it:-(3)

.... --- .-- / .-.. --- -. --. /

.. ... / - .... . /

. -. --. .-.. .. .. /

-.-. .-.. .- ... ...

U ••—

UNIFORM

1	U	☐	..-
2	EA	☐	.. -
3	IT	☐	.. .-
4	EET	☐	.. -

Writing Practice

U ••— Unit ..- -. .. -

U ••— Unit ..- -. .. -

Read Aloud

U — di di dah

U — di di dah

U — di di dah

U — di di dah

U — di di dah

Extract the letter "U" and circle it:-(3)

.-- --- ..- .-.. -.. / -.-- --- ..- /

.-". .-.. . . .- . ... . /

.-- .-. -.- . / -- . /

..- .-". / . .- - / --... .- --

A •— B C D —•• E • F G H •••• I •• J K L •—•• M —— N —• O ——— P Q R •—• S ••• T — U ••— V W X Y Z

○ Learned

C ■ — ■ • ■ • — ■ •

CHARLIE

1 C   □ -. -.

2 NN   □ - . - .

3 TETE   □ -.-.

4 TR   □ - .-.

Writing Practice

C ■ • ■ •   Cat -.-. .- -

Read Aloud

C   dah di dah dit

C   dah di dah dit

C   dah di dah dit

C   dah di dah dit

C   dah di dah dit

Extract the letter "C" and circle it:-(3)

.. / .-.. .. -.- . /

-.-. .... --- -.-. --- .-.. .- - . /

-.-. .- -.- .

A .-   C -.-. D -.. E .   H .... I ..   L .-.. M -- N -. O ---   R .-. S ... T - U ..-

©Learned

W · — —

WILLIAM

1 W   ☐ · —
2 AT   ☐ · — ·
3 EM   ☐ · — · ·
4 ETT   ☐ · —

Writing Practice

W · — —   What · — — · · · · · — · ·

Read Aloud

W   dit dah dah

W   dit dah dah

W   dit dah dah

W   dit dah dah

W   dit dah dah

Extract the letter "W" and circle it:-(2)

· · · ·   — — —   · — —   /   — · ·   — — —   /

— · — —   — — —   · · —   /   — — · ·   ·   —   /

— · — —   /   · — —   — — —   · — ·   — · —

Learned: A B C D E F G H I J K L M N O P Q R S T U V W X Y Z

A E M O T N S I R H D L U C W

Match Morse words with the correct Word

AIR	.− .. .−.
LAW	.−.. .− .−−
WIN	− .−. ...
TREE	.−− .. −.
DUCK	.−.. ..− −.−. −.−
LUCK	.−− −−− .−. −..
WORD	−.. ..− −.−. −.−

Match Morse words with the correct picture

... −.− .. .. −−− .−. ...	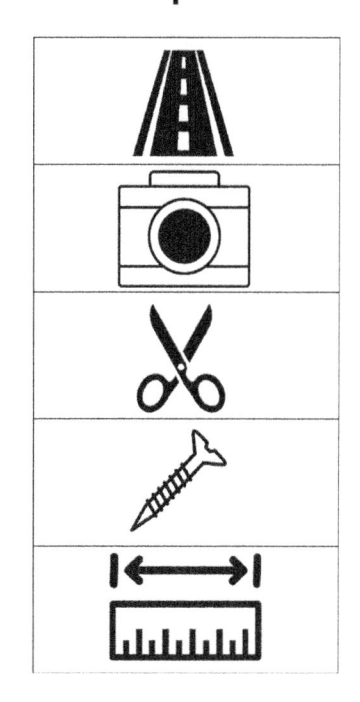
... −.−. .−. .−. . .−−	
.−. −−− .− −..	
.−. ..− .−.. . .−.	
−.−. .−− . −− . .−. .−ç	

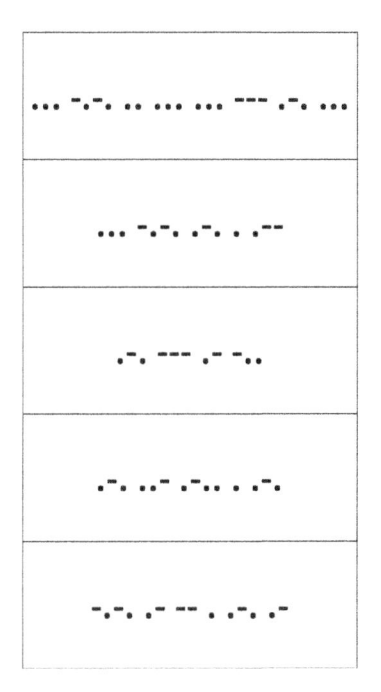

Puzzle: Decode the learned letters and extract 10 words 🔍

.-	-.-	.-.	..-.	-.-.	.-	.--	-.-.	...-	--.-	-	--
-.	.-..	.-..	.-.	..	-.-.	.	-	.-.	.	.	.-
..	-.-.	.-	-	--	.-.	.--	..	-.	.-..	--.-	-.
--	.--	.--	--	.	.-	-	.---	--.	.	.--	.-
.-	-.-.	.-.	.-	.-.	--	.-..	---	.--	-.	---	.-.
.-..	.-.	---	.-	-..	-...	..-	..-.	-.-.	...	.-.	-.-.
..-.	-.-.	.-	.-.	..	.--	---	---	-..	...-	-..	--..
-.	..	-.	.	.-	..	.-.	..	-.-.	.	.-.	-.--

Translate text to Morse code

AIR	
LENS	
WORD	
LAW	
ROAD	
CAW	

F ●●▬● FOXTROT

☐1 F   ☐ ●●▬●

☐2 ER   ☐ ●▬●●

☐3 ITE   ☐ ●● ▬●

☐4 IN   ☐ ●● ▬ ●

Writing Practice

F ●●▬● Fat ●●▬● ●▬▬

F ●●▬● Fat ●●▬● ●▬▬

Read Aloud

F di di dah dit

F di di dah dit

F di di dah dit

F di di dah dit

F di di dah dit

Extract the letter "F" and circle it:-(2)

●▬▬ ●●●● ●▬ ▬ / ▬●▬ ●● ▬● ▬●● /

▬▬▬ ●●▬● / ●●▬● ▬▬▬ ▬▬▬ ▬●● /

▬●● ▬▬▬ / ▬●▬▬ ▬▬▬ ●●▬ /

●▬●● ●● ▬●▬ ●

A ●▬ B ▬●●● C ▬●▬● D ▬●● E ● F ●●▬● G ▬▬● H ●●●● I ●● J ●▬▬▬ K ▬●▬ L ●▬●● M ▬▬ N ▬● O ▬▬▬ P ●▬▬● Q ▬▬●▬ R ●▬● S ●●● T ▬ U ●●▬ V ●●●▬ W ●▬▬ X ▬●●▬ Y ▬●▬▬ Z ▬▬●●

Y ■ — ● ■ —

YANKEE

1 Y	□ -.--
2 TAT	□ -.--
3 TETT	□ -.--
4 NM	□ -.--

Writing Practice

| Y ■—●■— | You -.-- --- ..- |
| Y | You |

Read Aloud

Y	dah di dah dah
Y	dah di dah dah
Y	dah di dah dah
Y	dah di dah dah
Y	dah di dah dah

Extract the letter "Y" and circle it:-(2)

-- -.-- / ..-. .-. .. . -. -.. /

.. ... / ..-. . .-. --- -- /

. --. -.-- .--. -

Learned A B C D E F G H I J K L M N O P Q R S T U V W X Y Z
.- -... -.-. -.. . ..-. --. ....--- -.- .-.. -- -. --- .--. --.- .-. ... - ..- ...- .-- -..- -.-- --..

P

PAPA

P · — — ·

1	P	☐	. — — .
2	R	☐	. — .
3	AN	☐	. — — .
4	EME	☐	. — — .

Writing Practice

P	· ■ ■ · Pay ·—— · · — — ·——
P	· ■ ■ · Pay ·—— · · — — ·——

Read Aloud

P — dit dah dah dit

P — dit dah dah dit

P — dit dah dah dit

P — dit dah dah dit

P — dit dah dah dit

Extract the letter "P" and circle it:-(2)

.. / .—— ——— .—. —.— /

.. —. / .— / .—— .— .—. . .—. /

..—. .— —.—. — ——— .—. —.——

A B C D E F G H I J K L M N O P Q R S T U V W X Y Z

G ▬ ▬ •

GOLF

① G		☐ -- .	
② TTE		☐ - - -.	
③ ME		☐ --.	
④ TN		☐ - -- .	

Writing Practice

G -- •   Go --. ---

Read Aloud

G   dah dah dit
G   dah dah dit
G   dah dah dit
G   dah dah dit
G   dah dah dit

Extract the letter "G" and circle it:-(2)

--. --- --- -.. /

-- --- .-. -. .. -. --.

A .- B C -.-. D -.. E . F ..-. G --. H .... I .. J K -.- L .-.. M -- N -. O --- P Q R .-. S ... T - U ..- W .-- X Y -.-- Z

B

B ▬ ● ● ●

BRAVO

1 B — □ -...

2 NI — □ -.. .

3 NEE — □ -... .

4 TIE — □ - ...

Writing Practice

B ▬●●● Bat -... .- -

B ▬●●● Fat -... .- -

Read Aloud

B dah di di dit

B dah di di dit

B dah di di dit

B dah di di dit

B dah di di dit

Extract the letter "F" and circle it:-(2)

.-- .... .- - / -.- .. -. -.. /

--- ..-. / ..-. --- --- -.. /

-.. --- / -.-- --- ..- /

.-.. .. -.- .

A B C D E F G H I J K L M N O P Q R S T U V W X Y Z

Learned

The Learned Letters

| A E M O T N S I R H D L U C W F Y P G B |

Match Morse words with the correct Word

FAMILY		--. .- -- .
GYM		.- .-- .-- .-.. .
PEACE		.-- . .- -.-. .
SISTER		..-. .- -- .. .-.. -.--
APPLE		... . - . .-.
GAME		--. -.-- --
GIRL		--. .. .-. .-..

Match Morse words with the correct picture

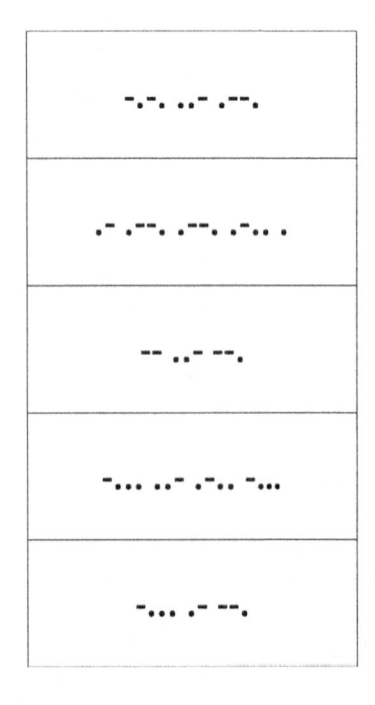

-.-. ..- .--.
.- .-- .-- . .-.. .
-- ..- --.
-... ..- .-.. -...
-... .-. --.

Puzzle: Decode the learned letters and extract 10 words 🔍

--.	..-.	.--.	.	.-	-.-.	.	..-	-..	-.-.	-..-	.-
.-	---	.-	.--.	.--.	.-..	.	...	..-.	..-	.---	..-.
--	---	--.	-.-.	.-..	..-	-...	..	.-	.--.	..-.	.-..
.	-..	..-	--.-	-..	---	--.	...	-	--.	.-..	.-
..-.	.-	--	..	.-.	-.--	-...	-	....	-.--	-.--	--
.--.	--	..-	--.	-...	---	-.--	.	.	--	.---	-...
--.	..	.-.	.-..	--	.-	.--.	.-.	.-.	--..	.-	....
..	.--.	-.-.	.-	-	...	--	.-	.-.	-	..-.	-...

Translate text to Morse code

MUG	
BAG	
CUP	
FLY	
Girl	
BROTHER	

V · · · —
VICTOR

① V		☐	...-
② IA		☐	. ..-
③ EIT		☐	.. .-
④ EU		☐	... -

Writing Practice

V · · · —	Vat ...- .- -
V	Vat

Read Aloud

V di di di dah
V di di di dah
V di di di dah
V di di di dah
V di di di dah

Extract the letter "B" and circle it:-(1)

.. / .- -- /

.-. . . .- -.. .. -. --. / .- /

-... --- --- -.-

A B C D E F G H I J K L M N O P Q R S T U V W X Y Z
.- -... -.-. -.. . ..-. --. ....--- -.- .-.. -- -. --- .--. --.- .-. ... - ..- ...- .-- -..- -.-- --..

K — • —

KILO

Match Letters to Morse Code

1 K   ☐ - . -

2 NT   ☐ - . -

3 TET   ☐ - . -

4 TA   ☐ - . -

Writing Practice

K   — • —   Kit   - . - . . -

K   — • —   Kit   - . - . . -

Read Aloud

K   dah di dah

K   dah di dah

K   dah di dah

K   dah di dah

K   dah di dah

Extract the letter "K" and circle it:-(2)

.. / .- .. -.. . /

-- -.-- / -... .. -.- . /

- --- / .-- --- .-. -.-

A	B	C	D	E	F	G	H	I		K	L	M	N	O	P		R	S	T	U	V	W	X	Y	
•—	—•••	—•—•	—••	•	••—•	——•	••••	••		—•—	•—••	——	—•	———	•——•		•—•	•••	—	••—	•••—	•——	—••—	—•——	

Learned

J ·▬▬▬

JULIETT

☐1 J		☐	.- --
☐2 EMT		☐	.--
☐3 EO		☐	. -- -
☐4 AM		☐	. ---

Writing Practice

J ·▬▬▬	Jar .--- .- .-.
J ·▬▬▬	Jar .--- .- .-.

Read Aloud

J di dah dah dah

J di dah dah dah

J di dah dah dah

J di dah dah dah

J di dah dah dah

Extract the letter "J" and circle it:-(1)

.. / . -. .--- --- -.-- /

- .... / .-- . . .- -.-. . . /

.- -. -.. / --.- ..- .. . .-

ⒺLearned A B C D E F G H I J K L M N O P Q R S T U V W X Y Z

X −··−

X-RAY

1 X		−··−
2 NA		−··−
3 TIT		− ··−
4 DT		−·· −

Writing Practice

X −··− Fox ··−· −−− −··−	
X −··− Fox ··−· −−− −··−	

Read Aloud

X dah di di dah

X dah di di dah

X dah di di dah

X dah di di dah

X dah di di dah

Extract the letter ''X'' and circle it:-(1)

·· / ·−− ·· −·· −·· / −−· −−− /

··· ···· −−− ·−−· ·−−· ·· −· −−· /

−·· · −··− − / ·−− ·· −·−

ABCDEFGHIJKLMNOPQRSTUVWXYZ

Q

Q ▬ ▬ ● ▬

QUEBEC

1	Q	☐	-- .-
2	MA	☐	--. -
3	GT	☐	- -.-
4	TK	☐	--.-

Writing Practice

Q   ▬▬●▬   Queen --.- .. . . -.

Q   ▬▬●▬   Queen --.- .. . . -.

Read Aloud

Q   dah dah di dah

Q   dah dah di dah

Q   dah dah di dah

Q   dah dah di dah

Q   dah dah di dah

Extract the letter "Q" and circle it:-(1)

.. / . -.- .--- --- -.-- /

- .... / .--. . . .- -.-. . /

.- -. -.. / --.- ..-. .. . -

A B C D E F G H I J K L M N O P Q R S T U V W X Y Z

Z

Z ▬ ▬ ● ●

ZULU

1	Z	☐	▬ ▬ . .
2	MI	☐	▬ ▬ . .
3	TNE	☐	▬ ▬ . .
4	GE	☐	▬ ▬ . .

Writing Practice

Z ▬▬●● Zebra --.. . -... .-. -.

Z Zebra

Read Aloud

Z — dah dah di dit

Z — dah dah di dit

Z — dah dah di dit

Z — dah dah di dit

Z — dah dah di dit

Extract the letter "Z" and circle it:-(1)

.. - / .. ... /

..-. .-. . . ▬▬.. .. -. ▬▬. /

▬ ▬▬▬ ▬.. .-. ▬.▬▬

A B C D E F G H I J K L M N O P Q R S T U V W X Y Z

The Learned Letters

| A E M O T N S I R H D L U C W F Y P G B V K J X Q Z |

Match Morse words with the correct Word

ANIMAL		.-.. .. --- -.
DUCK		.- -. .. -- .- .-..
LION		.--- ..- -. --. .-.. .
JUNGLE		-.. ..- -.-. -.-
JUMP		.--- ..- -- .--.
ZEBRA		--.- ..- .- .-.. .. - -.--
QUALITY		--.. . -... .-. .-

Match Morse words with the correct picture

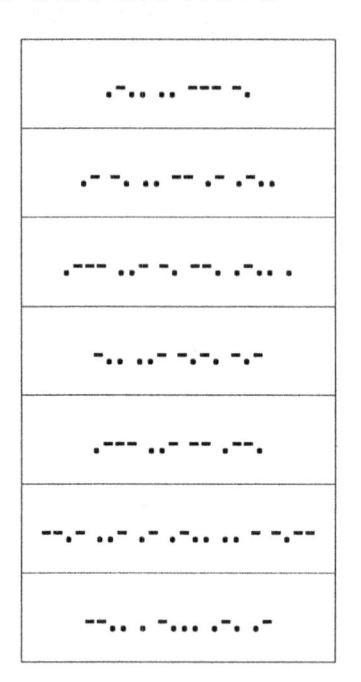

-.- .. - .		
.--. ..- --.. --.. .-.. .		
-... --- -..-		
--.- ..- . . -.		
...- .- -.. .		

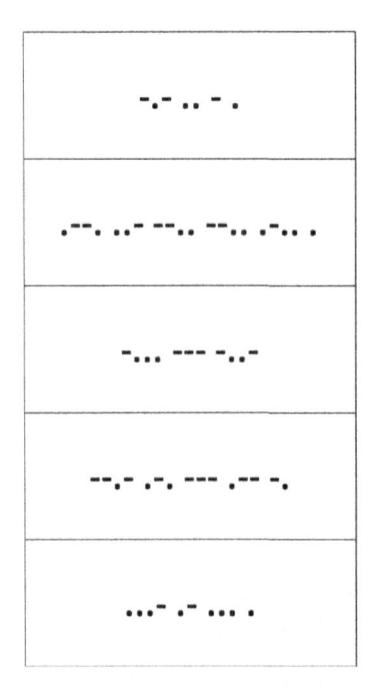
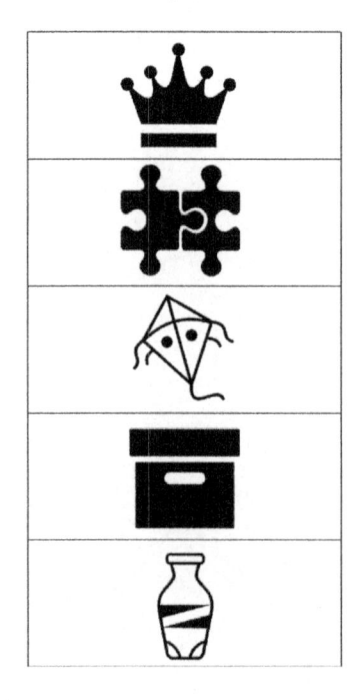

Puzzle: Decode the learned letters and extract 15 words 🔍

--..	...-	...-	-	.--	---	.-.	-..	--	..-	-.-	-...
---	.-	.-	--	---	.-.	...	.	.--.	-..	..	-...
---	...	-.	-..	-.--	.	.-..	.-..	---	.--	-	-.--
-..	.	-...	.-.	.-	..	-.-.	.	-...	-..-	.	..
...	--	.-	.-.	-	--.	.-.	.-.	.-.	.-..	.	-.
-...	---	-..-	.-	-.	-	-	.--	.-	.-.	-..	--.
..-.	---	-..-	-	.--	.--	..	-.	-.	.	.-.	-..
-..	...-	--	---	.-.	...	.	-.-.	---	-..	.	...-

Translate text to Morse code

KING	
PUZZLE	
FOX	
JAR	
QUIZ	
VASE	

The Learned Letters

A B C D E F G H I J K L M N O P Q R S T U V W X Y Z

Match Morse words with the correct Word

Word	Morse
Passenger	.--. --- .-- . .-.
Students	. -- . .-. --. . .-. .-. .-.
Power	..-. --- .-. --. . -
Emergency	.-- . .- - .... . .-.
Bee	.. -- .-. --- .-. . .-. -. -
Forget	.-. --- -.-. -.- . -
Confidential	- --- --. . - .... ..-.
Weather	-. --- - . -... --- --- -.-
Together	.--. .- ... . -. --. . .-.
Spring	-... ..
Rocket	-.-. --- -. ..-. .. -... -. - .. .- .-..
Danger	... . -.-. .-. . -
Important	-... . .- -. --. . .-.
Secret	... - ..- -.. . -. - ...
Notebook	... .--. .-. .. -. --.

Puzzle: Decode the learned letters and extract 15 words 🔍

A	B	C	D	E	F	G	H	I	J	K	L	M
·—	—···	—·—·	—··	·	··—·	——·	····	··	·———	—·—	·—··	——

N	O	P	Q	R	S	T	U	V	W	X	Y	Z
—·	———	·——·	——·—	·—·	···	—	··—	···—	·——	—··—	—·——	——··

Date   /   /   _____

A B C D E F G H I J K L M

N O P Q R S T U V W X Y Z

Date   /   /   _____

A   B   C   D   E   F   G   H   I   J   K   L   M
·— —··· —·—· —··   ·   ··—· ——· ···· ·· ·——— —·— ·—·· ——

N   O   P   Q   R   S   T   U   V   W   X   Y   Z
—· ——— ·——· ——·— ·—· ···   —   ··— ···— ·—— —··— —·—— ——··

A	B	C	D	E	F	G	H	I	J	K	L	M
•—	—•••	—•—•	—••	•	••—•	——•	••••	••	•———	—•—	•—••	——

N	O	P	Q	R	S	T	U	V	W	X	Y	Z
—•	———	•——•	——•—	•—•	•••	—	••—	•••—	•——	—••—	—•——	——••

MORSE CODE

Learn and practice Morse code in a different way by using your phone

Scan QR code
to learn and
practice
Morse Code

"The secret of getting ahead is getting started"
– Mark Twain

1	2	3	4	5
·−−−−	··−−−	···−−	····−	·····

6	7	8	9	0
−····	−−···	−−−··	−−−−·	−−−−−

(		@		)
−·−−·		·−−·−·		−·−−·−

_	?	!	,	,
·−···−	··−−··	−·−·−−	·−·−·−	−−··−−

+	-	/	=	:
·−·−·	−····−	−··−·	−···−	−−−···

1 •----

Match Letters to Morse Code

☐1 1   ☐ •----

☐2 9   ☐ •- ---

☐3 AO   ☐ • •• ••

☐4 Ell   ☐ ----•

Writing Practice	Read Aloud

Writing Practice

1 •----   19 •---- ----•

1   19

Read Aloud

1   di dah dah dah dah

1   di dah dah dah dah

1   di dah dah dah dah

1   di dah dah dah dah

1   di dah dah dah dah

Extract the Number "1" and circle it:-(6)

•---- ----• •---- ----• •----

•---- ••--- •••-- ••••-

----- •---- •---- ••--- -••••

2 ·· ▬ ▬ ▬

1 2		·· ---
2 8		··---
3 IO		---··
4 IMT		·· --- -

Writing Practice

| 2 ·· ▬ ▬ ▬ | 28 ··--- ---·· |
| 2 ·· ▬ ▬ ▬ | 28 ··--- ---·· |

Read Aloud

2   *di di dah dah dah*

2   *di di dah dah dah*

2   *di di dah dah dah*

2   *di di dah dah dah*

2   *di di dah dah dah*

Extract the Number "2" and circle it:-(4)

··--- ---·· ··--- ---·· ··---

·---- ··--- ···-- ····-- ·····

····· ····- -···· ···-- ·····

3 ···━━

☐ 3   ☐ ·· ·- -

☐ 7   ☐ --···

☐ SM   ☐ ···--

☐ IAT   ☐ ··· --

Writing Practice

3 ···━━   37   ···-- --···

Read Aloud

3   di di di dah dah

3   di di di dah dah

3   di di di dah dah

3   di di di dah dah

3   di di di dah dah

Extract the Number "3" and circle it:-(3)

·---- ··--- ···-- ····-

····- ····· -···· --···

···-- ---·· --··· --··· ···--

0   1   2   3   4   5   6   7   8   9

4 •••• ▬

1 4	☐	-••••
2 6	☐	•••• -
3 HT	☐	•••• -
4 IIT	☐	••••-

Writing Practice

4 •••• ▬   46 ••••- -••••

4 •••• ▬   46 ••••- -••••

Read Aloud

4 di di di di dah

4 di di di di dah

4 di di di di dah

4 di di di di dah

4 di di di di dah

Extract the Number "4" and circle it:-(2)

-•••• •▬▬▬ ••▬▬▬ •••▬▬ ••••▬

••••▬ ••••• ▬•••• ▬▬•••

▬▬•• ▬▬▬▬• ▬▬▬▬▬

5 ● ● ● ● ●

1	5		..... .
2	0		-----
3	HE		.. .. .
4	IIE		.....

Writing Practice

5 ● ● ● ● ●   50   ..... -----

5 ● ● ● ● ●   50   ..... -----

Read Aloud

5   di di di di dit

5   di di di di dit

5   di di di di dit

5   di di di di dit

5   di di di di dit

Extract the Number "5" and circle it:-(3)

●----- ..--- ...-- ....- .....

..... -.... --... ---.. ----.

-----. ...-- ...-- ....---

0   1   2   3   4   5   6   7   8   9

6 ▬ • • • •

1	6	☐	-....
2	4	☐	....-
3	TH	☐	-....
4	NEI	☐	-....

Writing Practice

6 ▬ • • • • 64 ....- -....

6 ▬ • • • • 64 ....- -....

Read Aloud

6 dah di di di dit

6 dah di di di dit

6 dah di di di dit

6 dah di di di dit

6 dah di di di dit

Extract the Number "6" and circle it:-(3)

....- ..... -... ....- -----

--.. --... ....- ....- .----

-.... ...-- .. .--- ....- -....

0 1 2 3 4 5 6 7 8 9

7 ▬ ▬ ● ● ●

1. 7   □ -....
2. 3   □ ....-
3. TH   □ -....
4. NEI   □ -. . ..

Writing Practice

7 ▬ ▬ ● ● ●   73   --... ...--

7 ▬ ▬ ● ● ●   73   --... ...--

Read Aloud

7   dah dah di di dit
7   dah dah di di dit
7   dah dah di di dit
7   dah dah di di dit
7   dah dah di di dit

Extract the Number "7" and circle it:-(2)

●●●●- ●●●●● -●●●● --●●● ---●●

--●●● ●-▬▬▬ ●-▬▬▬ ●●●●- ●●●--

-●●●● ●●●-- ●●-▬▬ ●●●●-

0   1   2   3   4   5   6   7   8   9
●▬ ▬▬ ●●▬▬ ●●●▬ ●●●●▬ ●●●●● ▬●●●● ▬▬●●●

8 ▬ ▬ ▬ • •

1 8   ☐ ---..

2 2   ☐ ..---

3 OI   ☐ -- -..

4 MTI   ☐ --- ..

Writing Practice		Read Aloud
8 ▬▬▬••	82 ---.. ..---	8 dah dah dah di dit
8	82	8 dah dah dah di dit
		8 dah dah dah di dit
		8 dah dah dah di dit
		8 dah dah dah di dit

Extract the Number "8" and circle it:-(3)

.---- ..--- ...-- ..--- ---..

..... --... ....- -.... ---..

..--- ...-- .---- ----. ....-

0   1   2   3   4   5   6   7   8   9

.---- ..--- ...-- ....- ..... -.... --... ---..

9 ▄▄ ▄▄ ▄▄ ▄▄ ●

☐1	9	☐	.----
☐2	I	☐	.. ... .
☐3	IIE	☐	--- .-
☐4	OA	☐	----.

Writing Practice

9 ▄▄ ▄▄ ▄▄ ● 91 .---- ----.

9 ▄▄ ▄▄ ▄▄ ● 91 .---- ----.

Read Aloud

9   dah dah dah dah dit

9   dah dah dah dah dit

9   dah dah dah dah dit

9   dah dah dah dah dit

9   dah dah dah dah dit

Extract the Number "9" and circle it:-(4)

.---- ----. .---- ----.

.---- ----. ----.

.---- ..--- -.... ....- .....

1 .----   2 ..---   3 ...--   4 ....-   5 .....   6 -....   7 --...   8 ---..   9 ----.

0 ▬ ▬ ▬ ▬ ▬

☐ 0	☐ -- ---
☐ 5	☐ -----
☐ MO	☐ -- -- -
☐ MMT	☐ .....

Writing Practice

0 ▬ ▬ ▬ ▬ ▬   05   ----- .....

0 -----   05

Read Aloud

0   dah dah dah dah dah

0   dah dah dah dah dah

0   dah dah dah dah dah

0   dah dah dah dah dah

0   dah dah dah dah dah

Extract the Number "0" and circle it:-(4)

----- .---- ..--- ...-- ....-

...-- ..--- ..... ----- ..---

....- ----- .---- ..--- -----

Learned   0   1   2   3   4   5   6   7   8   9

Date   /   /   _____

0   1   2   3   4   5   6   7   8   9

0   1   2   3   4   5   6   7   8   9
▬▬▬▬▬ ·▬▬▬▬ ··▬▬▬ ···▬▬ ····▬ ····· ▬···· ▬▬··· ▬▬▬·· ▬▬▬▬·

Date / /

0　1　2　3　4　5　6　7　8　9

0   1   2   3   4   5   6   7   8   9

0	1	2	3	4	5	6	7	8	9
▬▬▬▬▬	•▬▬▬▬	••▬▬▬	•••▬▬	••••▬	•••••	▬••••	▬▬•••	▬▬▬••	▬▬▬▬•

A B C D E F G H I

J K L M N O P Q R

S T U V W X Y Z

1 .----

2 ..---

3 ...--

4 ....-

5 .....

6 -....

7 --...

8 ---..

9 ----.

0 -----

Take it easy

- .- -.- . / .. - / . .- ... -.--   *(Translation)*

I am good

How are you

How you doing

A	B	C	D	E	F	G	H	I	J	K	L	M
•—	—•••	—•—•	—••	•	••—•	——•	••••	••	•———	—•—	•—••	——

N	O	P	Q	R	S	T	U	V	W	X	Y	Z
—•	———	•——•	——•—	•—•	•••	—	••—	•••—	•——	—••—	—•——	——••

Would you mind closing the door

How much does this cost

A	B	C	D	E	F	G	H	I	J	K	L	M
·—	—···	—·—·	—··	·	··—·	——·	····	··	·———	—·—	·—··	——

N	O	P	Q	R	S	T	U	V	W	X	Y	Z
—·	———	·——·	——·—	·—·	···	—	··—	···—	·——	—··—	—·——	——··

I live close to my family

I brush my teeth every day

A	B	C	D	E	F	G	H	I	J	K	L	M
•━	━•••	━•━•	━••	•	••━•	━━•	••••	••	•━━━	━•━	•━••	━━

N	O	P	Q	R	S	T	U	V	W	X	Y	Z
━•	━━━	•━━•	━━•━	•━•	•••	━	••━	•••━	•━━	━••━	━•━━	━━••

I am learning Morse Code

I am reading a book

A	B	C	D	E	F	G	H	I	J	K	L	M
·—	—···	—·—·	—··	·	··—·	——·	····	··	·———	—·—	·—··	——

N	O	P	Q	R	S	T	U	V	W	X	Y	Z
—·	———	·——·	——·—	·—·	···	—	··—	···—	·——	—··—	—·——	——··

Save our souls SOS

SOS

Help

A	B	C	D	E	F	G	H	I	J	K	L	M
•—	—•••	—•—•	—••	•	••—•	——•	••••	••	•———	—•—	•—••	——

N	O	P	Q	R	S	T	U	V	W	X	Y	Z
—•	———	•——•	——•—	•—•	•••	—	••—	•••—	•——	—••—	—•——	——••

\-\-. \-\-\- \-\-\- \-.. / .\- ..\-. \-. . .\-. \-. \-\-\- \-\-\- \-.

Good Afternoon — Decoding

.... \-\-\- .\-\- / .\- \-. . / \-.\-\- \-\-\- ..\- / \-.. \-\-\- .. \-. \-\-.

A	B	C	D	E	F	G	H	I	J	K	L	M
.\-	\-...	\-.\-.	\-..	.	..\-.	\-\-.	....	..	.\-\-\-	\-.\-	.\-..	\-\-

N	O	P	Q	R	S	T	U	V	W	X	Y	Z
\-.	\-\-\-	.\-\-.	\-\-.\-	.\-.	...	\-	..\-	...\-	.\-\-	\-..\-	\-.\-\-	\-\-..

.. / .-- .- -.- . / ..- .--. / .- - / --... .- --

.. / --. --- / - --- / .-- --- .-. -.- /

. ...- . .-. -.-- / -.. .- -.--

A .- B -... C -.-. D -.. E . F ..-. G --. H .... I .. J .--- K -.- L .-.. M --

N -. O --- P .--. Q --.- R .-. S ... T - U ..- V ...- W .-- X -..- Y -.-- Z --..

Solution: I GO TO WORK EVERY DAY

.. / .-.. .. -.- . / - --- / ... .--. . -. -.. / - .. -- . /

.-- .. - .... / -- -.-- / ..-. .- -- .. .-.. -.-- / .- -. -.. /

..-. .-. .. . -. -.. ... / .. -. / -- -.-- / ..-. .-. . . /

- .. -- .

A	B	C	D	E	F	G	H	I	J	K	L	M
.-	-...	-.-.	-..	.	..-.	--.	....	..	.---	-.-	.-..	--

N	O	P	Q	R	S	T	U	V	W	X	Y	Z
-.	---	.--.	--.-	.-.	...	-	..-	...-	.--	-..-	-.--	--..

.... --- .-- / .- .-. . / -.-- --- ..- / -.. --- .. -. --.

A	B	C	D	E	F	G	H	I	J	K	L	M
•—	—•••	—•—•	—••	•	••—•	——•	••••	••	•———	—•—	•—••	——

N	O	P	Q	R	S	T	U	V	W	X	Y	Z
—•	———	•——•	——•—	•—•	•••	—	••—	•••—	•——	—••—	—•——	——••

Decode Morse code to text

.... --- .-- / .- .-. . / -.-- --- ..- / -.. --- .. -. --. ..

.-.. --- --- -.- / ..-. --- .-. .-- .- .-. -.. / - ---

... . -. --. / -.-- --- ..- / .- --. .- .. -.

A .- B -... C -.-. D -.. E . F ..-. G --. H .... I .. J .--- K -.- L .-.. M --

N -. O --- P .--. Q --.- R .-. S ... T - U ..- V ...- W .-- X -..- Y -.-- Z --..

..--- --.. / .-. . ..- -... .-.. .. ..-.

A	B	C	D	E	F	G	H	I	J	K	L	M
.-	-...	-.-.	-..	.	..-.	--.	....	..	.---	-.-	.-..	--

N	O	P	Q	R	S	T	U	V	W	X	Y	Z
-.	---	.--.	--.-	.-.	...	-	..-	...-	.--	-..-	-.--	--..

SOS

HELP

A	B	C	D	E	F	G	H	I	J	K	L	M
.-	-...	-.-.	-..	.	..-.	--.	....	..	.---	-.-	.-..	--

N	O	P	Q	R	S	T	U	V	W	X	Y	Z
-.	---	.--.	--.-	.-.	...	-	..-	...-	.--	-..-	-.--	--..

MORSE CODE

Learn and practice Morse code in a different way by using your phone

Scan QR code to learn and practice Morse Code

"The secret of getting ahead is getting started"
– Mark Twain

1 •－－－－

2 ••－－－

3 •••－－

4 ••••－

5 •••••

6 －••••

7 －－•••

8 －－－••

9 －－－－•

0 －－－－－

(－•－－•

@ •－－•－•

) －•－－•－

. •－•－•－

? ••－－••

! －•－•－－

' •－－－－•

, －－••－－

+ •－•－•

- －••••－

/ －••－•

= －•••－

: －－－•••

Match Letters to Morse Code

1 =   □ -.. .-

2 DA   □ -...-

3 TST   □ -. ... -

4 NEET   □ - ... -

Writing Practice		Read Aloud

=   ■ ● ● ● ■   =5   -...- .....

=   *dah di di di dah*

=   *dah di di di dah*

=   *dah di di di dah*

=   *dah di di di dah*

=   *dah di di di dah*

Extract the Symbol "=" and circle it:-(1)

-.. .- / ...-- ..--- ....--

-...- ...-- ..---- ....--

/ -.. .- /

Match Letters to Morse Code

☐1 - ☐ -.. ..-

☐2 DU ☐ -....-

☐3 THT ☐ -. .. . -

☐4 NIET ☐ - .... -

Writing Practice

-= -....- -...-

Read Aloud

- dah di di di di dah

- dah di di di di dah

- dah di di di di dah

- dah di di di di dah

- dah di di di di dah

Extract the Symbol "-" and circle it:-(1)

..... -....- ...-- -...- ..---

-...- ...-- ..---- ....--

/ -.. .- /

— Minus

Match Letters to Morse Code

+ ● ━ ● ━ ●

1 +   □ .-. -.

2 RN   □ . - -.

3 RTE   □ .-. -.

4 ETR   □ .-.-.

Writing Practice		Read Aloud

+ ● - ● - ● | +- .-.- -...- | + di dah di dah dit

+ | +- | + di dah di dah dit

+ di dah di dah dit

+ di dah di dah dit

+ di dah di dah dit

Extract the Symbol "+" and circle it:-(2)

●●●● -●●●- ●●●-- -●●●- ---●●

●●●● ●-●-● ●●●-- -●●●- ---●●

●----- ●●●--- ●●●-- / ●-●-●

= - + . / : ? ! ' ,

1. .

2. RK

3. RTA

4. ETAA

☐ .-. -.-

☐ .-. - . -

☐ .-.-.-

☐ . - .- .-

Writing Practice		Read Aloud

Read Aloud:

• di dah di dah di dah

• di dah di dah di dah

• di dah di dah di dah

• di dah di dah di dah

• di dah di dah di dah

Extract the Symbol "." and circle it:-(2)

..... -....- ...-- -...- ---..

..... .-.-. .-.-. -...- ---..

.-.-. ..--- ...-- / .-.-.

Learned

= - + . / : ? ! , ,

	Match Letters to Morse Code

/   ▬ ● ● ▬ ●

Match Letters to Morse Code

1. /   ☐ -..-.

2. NR   ☐ -.. -.

3. NEN   ☐ -. -.

4. TITE   ☐ - .. - .

Writing Practice		Read Aloud
/ ▬●●▬● /. -..-. .-.-.		/ dah di di dah dit
		/ dah di di dah dit
		/ dah di di dah dit
		/ dah di di dah dit
		/ dah di di dah dit

Extract the Symbol "/" and circle it:-(1)

..... -...- ...-- -...- ---..

.---- ----- -..-. ..... -..-

..--- / .---- ..--- ...-- / .-.-.

=   -   +   .   /

-....-   -....-   .-.-.   .-.-.-   -..-.

Match Letters to Morse Code

- 1 :
- 2 OS
- 3 MNI
- 4 TMEI

- ---...
- --- ...
- - -- . ..
- -- -. ..

Writing Practice		Read Aloud

: ▬▬▬... :/ ---... -..-.

dah dah dah di di dit

dah dah dah di di dit

dah dah dah di di dit

dah dah dah di di dit

dah dah dah di di dit

Extract the Symbol ":" and circle it:-(2)

.---- ..--- ...-- ....----

....- --... ..... -....

.---- ..--- ...-- / ---...

○Learned

= - + . / :

? ! , ,

? •‒‒••

Match Letters to Morse Code

☐1 ?   ☐ •• ‒‒ ••

☐2 UD   ☐ ••‒ ‒••

☐3 IMI   ☐ • •‒ ‒ ••

☐4 EATI   ☐ ••‒‒••

Writing Practice

? •‒‒••   ?. ••‒•• •‒•‒

Read Aloud

? di di dah dah di dit

? di di dah dah di dit

? di di dah dah di dit

? di di dah dah di dit

? di di dah dah di dit

Extract the Symbol "?" and circle it:-(1)

•‒‒ •••• •‒ ‒ / •• ••• / ‒ •••• •‒ ‒

••‒‒•• / •‒‒ •••• •‒ ‒ / •• •••

= ‒••••   - ‒•••   + •‒•‒•   •. •‒•‒•‒   / ‒••‒•   : ‒‒‒•••   ? ••‒‒••   !   '

Match Letters to Morse Code

! `-.-.--`

1	!	☐	`-.-.--`
2	KW	☐	`-.-.--`
3	TRM	☐	`-.-.--`
4	TENM	☐	`-.-.--`

Writing Practice

! `-.-.--` !? `-.-.-- ..--..`

! `-.-.--` !? `-.-.-- ..--..`

Read Aloud

! dah di dah di dah dah

! dah di dah di dah dah

! dah di dah di dah dah

! dah di dah di dah dah

! dah di dah di dah dah

Extract the Symbol "!" and circle it:-(1)

`..-- ---- ..--- ..--- .--.-.`

`..--- --.. /`

`.--. ..-. -... .-.. .. ... .... . .-.`

`-.-.--`

`=` `·····-` `-` `-·····` `+` `·-·-·` `.` `·-·-·-` `/` `-··-·` `:` `---···` `?` `··--··` `!` `-·-·--` `'` `·----·` `,` `--··--`

,

● ▬ ▬ ▬ ●

1	'		□	.‐‐‐‐.
2	WG		□	.‐ ‐ ‐.
3	AMN		□	.‐ ‐ ‐.
4	AMTE		□	.‐‐ ‐‐.

Writing Practice

,
● ▬ ▬ ▬ ●   "   .‐‐‐‐. .‐‐‐‐.

Read Aloud

| di dah dah dah dah dit

| di dah dah dah dah dit

| di dah dah dah dah dit

| di dah dah dah dah dit

| di dah dah dah dah dit

Extract the Symbol " ' " and circle it:-(1)

..▬▬ ▬▬▬▬▬ ..▬▬ ..▬▬ .▬▬.▬.

..▬▬ ▬▬.. / .‐‐‐‐.

.▬▬. ..▬ ▬... .▬. . ...▬.

.‐‐‐‐.

=	-	+	.	/	:	?	!	'	
▬...▬	▬....▬	.▬.▬.	.▬.▬.▬	▬..▬.	▬▬...	..▬▬..	▬.▬.▬▬	.▬▬▬▬.	

	Match Letters to Morse Code

■ ■ ● ● ■ ■

,

☐ ,	☐ --. . - -
② ZM	☐ --.. --
③ MIM	☐ -- .. --
④ GETT	☐ --..--

Writing Practice		Read Aloud

, ■ ■ ● ● ■ ■ , --..-- .----. , *dah dah di di dah dah*

, ■ ■ ● ● ■ ■ , --..-- .----. , *dah dah di di dah dah*

, *dah dah di di dah dah*

, *dah dah di di dah dah*

, *dah dah di di dah dah*

Extract the Symbol " , " and circle it:-(2)

..-- ----- ..-- ..-- --..--

..-- --.. / .----.

.--. ..- -... .-.. .. ..-.

--..--

= ..--- - -....- + .-.-. . .-.-.- / -..-. : ---... ? ..--.. ! -.-.-- ' .----. , --..--

Match Letters to Morse Code

☐1 (		☐ -.--.
☐2 KN		☐ -.- - .
☐3 KTE		☐ -.- - .
☐4 TEME		☐ - . -- .

Writing Practice

(■•■■• (= -.--. -...-

(dah di dah dah dit

(dah di dah dah dit

(dah di dah dah dit

(dah di dah dah dit

(dah di dah dah dit

Read Aloud

Extract the Symbol "(" and circle it:-(1)

-.--. .---- ..--- ...-- .....

----. ....- -....

--... -.... ...-- .----

Match Letters to Morse Code

) **—·——·—**

☐1)		☐ -.--.-
☐2 KK		☐ -.- -.-
☐3 TPT		☐ - . -- .-
☐4 TEMA		☐ - .--. -

Writing Practice		Read Aloud

) **—·——·—**)= -.--.- -...-) dah di dah dah di dah

))=) dah di dah dah di dah

) dah di dah dah di dah

) dah di dah dah di dah

) dah di dah dah di dah

Extract the Symbol ")" and circle it:-(2)

-.--. - .---- ..--- ...-- .....

-.--.- ....- -....

---.. -.... ...-- -.--.-

⊘ Learned

(@)
-.--. .-.-.- -.--.-

@ ·−−·−·

☐ @	☐ ·−−· −·
☐ PN	☐ ·−·−·
☐ ANN	☐ ·−−·−·
☐ ATEN	☐ ·−−·−·

① @
② PN
③ ANN
④ ATEN

Writing Practice	Read Aloud

@ ·−−·−· @= ·−−·−· −···−

@ *di dah dah di dah dit*

@ *di dah dah di dah dit*

@ *di dah dah di dah dit*

@ *di dah dah di dah dit*

@ *di dah dah di dah dit*

Extract the Symbol "@" and circle it:-(1)

··−−− −−−−− ··−−− ··−−−

·−−·−· ··−−− −−·· /

·−−· ··− −··· ·−·· ·· ··· ···· · ·−·

(@)
·−·−· ·−−·−· −·−·−·

A B C D E F G H I

J K L M N O P Q R

S T U V W X Y Z

1	2	3	4	5
.————	..———	...——	....—	.....

6	7	8	9	0
—....	——...	———..	————.	—————

(	@	)
—.——.	.——.—.	—.——.—

.	?	!	'	,
.—.—.—	..——..	—.—.——	.————.	——..——

+	-	/	=	:
.—.—.	—....—	—..—.	—...—	——......

The sun is a huge ball of gases. It has a diameter of 1,392,000 km. It is so huge that it can hold millions of planets inside it. The Sun is mainly made up of hydrogen and helium gas. The surface of the Sun is known as the photosphere. The photosphere is surrounded by a thin layer of gas known as the chromosphere. Without the Sun, there would be no life on Earth. There would be no plants, no animals, and no human beings. As all the living things on Earth get their energy from the Sun for their survival.

A	B	C	D	E	F	G	H	I	J	K	L	M
·—	—···	—·—·	—··	·	··—·	——·	····	··	·———	—·—	·—··	——

N	O	P	Q	R	S	T	U	V	W	X	Y	Z
—·	———	·——·	——·—	·—·	···	—	··—	···—	·——	—··—	—·——	——··

Family is the place where you learn your first lesson in life. Your family members are the only assets that will remain with you forever. Whatever the circumstances, family members are always there for each other to support us. Good values and good morals are always taught in a family. In the family, we are prepared to respect our elders and love younger ones. We learn lessons consistently from our family, about honesty, dependability, kindness, and so on. Family always provides us with a sensation of so much love and care.

A	B	C	D	E	F	G	H	I	J	K	L	M
.—	—...	—.—.	—..	.	..—.	——.	....	..	.———	—.—	.—..	——

N	O	P	Q	R	S	T	U	V	W	X	Y	Z
—.	———	.——.	——.—	.—.	...	—	..—	...—	.——	—..—	—.——	——..

India is an agricultural country. Most of the people live in villages and are farmers. They grow cereals, pulses, vegetables, and fruits. The farmers lead a tough life. They get up early in the morning and go to the fields. They stay and work on the farm late till evening. The farmers usually live in kuchcha houses. Though they work hard they remain poor. Farmers eat simple food; wear simple clothes and rear animals like cows, buffaloes, and oxen. Without them, there would be no cereals for us to eat. They play an important role in the growth and economy of a country.

A	B	C	D	E	F	G	H	I	J	K	L	M
.—	—...	—.—.	—..	.	..—.	——.	....	..	.———	—.—	.—..	——

N	O	P	Q	R	S	T	U	V	W	X	Y	Z
—.	———	.——.	——.—	.—.	...	—	..—	...—	.——	—..—	—.——	——..

Decode Morse code to text

A B C D E F G H I J K L M N O P Q R S T U V W X Y Z

-.... . / ... --- -.. . -. / -. / ... -.. ... - . -- / -.. --- - . ... - ... / ---

..-. / - .... . / ... ..-. -. / -- --- --- -. / .- . -. / .-- .-. . -. . - . .-.-

/ .. - / .- . .-. . ... --- / -.-. --- -. . ... - ... / --- ..-. / -.-. --- -- . - ...

-- ..-- / -- . - . -- .-. --- .. -. . ... --..-- / .- . -.. / .- ... - . . --- .. -..

... ..-.-.- / - .... / ... ..- .- -. / .. ... / - .... . / .-.. .- -. . --. ... - / -- . --

-.... ..-. / .- - ..-. . - . . ... . / .. -- -.. .- . ... - . . -- .-.-.- / ..

-. / --- .- . .-. / --- ..-. / -.... - . -. . . . / ..-. .-. --- -- / - .... .

/ ... ..-. - -. --..-- / - .... . / .-- . - -.-. . ... / - .-. .. . .. .-. -.

-.-- --..-- / ... - . -. . .- ... --..-- / .- . -. . ... --..-- /

.--- ..- -. .- . -. .- . .--.-- / ... - . .. .- ... --..-- / .- . . ... --..--

--..-- / -. .. ... .- -. - . - ..- . -.- / .- -. . -. . -. . .- / - ....

/ -... .- -. . - . -. . .-. / -- .- . -. . . . / - .... . / .. .... / .- -

/ - .... . . / .-. -. . . -. . / --- ..- -. / .- .... . / ... -.-- ... -

. .-- / .- . -. . -.. / - .... . / .-- -. . .-. .-.-. / ... - . -.-- ... -

... --..-- / -.-. -. --- --- -- - . ... --..-- / .- . -. . -... / -- . . . .

.- .-.-.- / .- - .... . . / .-.-. . -. . . / --- -.. . / - .... . / ... ... / .-- -

/ - .... . . / .-.-. -. . . . -. . .- . / --- -. . -.. . / - .... . / -- . . . . .- . . .- .- -. ... / ... . .- . . .. /

.-.... ...- --- -. . -.. ... .- . / .- - -. --- ..- - -. . -.. / .. . .-.-

A B C D E F G H I J K L M N O P Q R S T U V W X Y Z

The **2Z Publisher** team Hopes that you enjoyed this book.

We would really love to get your **review** and positive feedback on Amazon.

2Z Publisher

..--- --..   .--. ..- -... .-.. .. ..-.

Write your memoirs in a secret way using one of our products as (Notebook: Morse Code Secret Notebook) form 2Z Publisher.

Practice Morse Code using one of our practice sheets as (Morse Code Practice Notebook).

2Z Publisher has many books about Morse code and other coding methods.

Visit our Page on AMAZON by scanning the QR code, using the next URL, or searching with "2Z Publisher" in the Amazon search engine for more info.

amazon.com/author/2zpublisher

What's Your Story?
Tell us your story about learning Morse code as you can inspire others to learn.
We would really love to get your **review** and feedback on Amazon.

Thanks for your trust in our products.

2Z Publisher
..--- --..

Printed in Great Britain
by Amazon

54008965R00071